DOING WORK WITH SIMPLE MACHINES

WORKING WITH
INCLINED
PLANES

RONALD MACHUT

PowerKiDS
press.
New York

Published in 2020 by The Rosen Publishing Group, Inc.
29 East 21st Street, New York, NY 10010

Editor: Elizabeth Krajnik
Book Design: Reann Nye

Photo Credits: Cover Christina Richards/Shutterstock.com; p. 7 BCFC/Shutterstock.com; p. 9 Rusla Ruseyn/Shutterstock.com; p. 10 romakoma/Shutterstock.com; p. 11 robert cicchetti/Shutterstock.com; p. 13 Atstock Productions/Shutterstock.com; p. 14 Waj/Shutterstock.com; p. 15 Elena Elisseeva/Shutterstock.com; p. 17 (top) Dmytro Budnik/Shutterstock.com; p. 17 (bottom) alexandrum79/Shutterstock.com; p. 18 riopatuca/Shutterstock.com; p. 19 Jason Kolenda/Shutterstock.com; p. 21 (top) Universal History Archive/Universal Images Group/Getty Images; p. 21 (bottom) LianeM/Shutterstock.com; p. 22 zefart/Shutterstock.com.

Library of Congress Cataloging-in-Publication Data

Names: Machut, Ronald, author.
Title: Working with inclined planes / Ronald Machut.
Description: New York : PowerKids Press, [2020] | Series: Doing work with simple machines | Includes index.
Identifiers: LCCN 2018025104| ISBN 9781538343609 (library bound) | ISBN 9781538345269 (pbk.) | ISBN 9781538345276 (6 pack)
Subjects: LCSH: Inclined planes–Juvenile literature.
Classification: LCC TJ1428 .M33 2019 | DDC 621.8–dc23
LC record available at https://lccn.loc.gov/2018025104

Manufactured in the United States of America

CPSIA Compliance Information: Batch #CSPK19: For Further Information contact Rosen Publishing, New York, New York at 1-800-237-9932

CONTENTS

A SIMPLE MACHINE

An inclined plane is a simple machine. A simple machine is a **device** with few or no moving parts that is used to modify, or change, motion and force to do work. The inclined plane is one of six simple machines.

An inclined plane is a surface with one end raised above the other. An inclined plane is used to move an object from a lower place to a higher place or from a higher place to a lower place. We push the object along the surface of the plane.

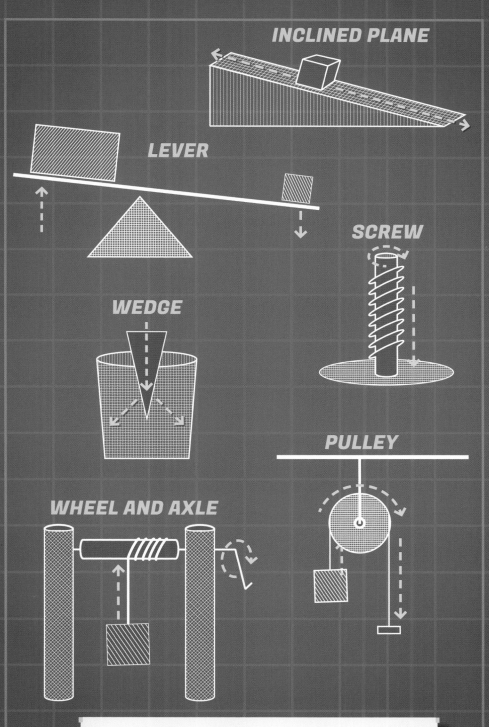

INCLINED PLANE

LEVER

SCREW

WEDGE

PULLEY

WHEEL AND AXLE

The other simple machines are the pulley, the lever, the wedge, the wheel and axle, and the screw.

5

WHAT ARE INCLINED PLANES MADE OF?

An inclined plane is just a sloped surface. Its measurements form an imaginary triangle. People or objects travel along the sloped surface of the inclined plane. This surface rises at an angle from the **horizontal**, or the flat base of the triangle. The height the plane covers is the **vertical** distance from the base to the top of the sloped surface.

Inclined planes can be man-made or they can occur in nature. They might be part of a **permanent** thing, such as a road, or be a **temporary** thing, such as a ramp used to load heavy objects into a truck.

Sliding a heavy box up this ramp is easier than lifting it.

MECHANICAL ADVANTAGE

←-------------------------------------→

Mechanical advantage is how much a machine changes the force you have to put into a job. Using a simple machine makes a job easier. The person using the machine has to put less force into the job.

Inclined planes provide a mechanical advantage when moving heavy objects. Take a moving ramp, for example. The mechanical advantage of a long ramp is greater than the mechanical advantage of a shorter, steeper ramp.

Imagine yourself standing at the bottom of a tall mountain. Would you rather climb straight up the steep side of the mountain or walk up a gradual, winding road to get to the top?

9

An inclined plane's mechanical advantage is **determined** by the slope's horizontal distance and the height of the slope. These measurements tell you the angle of incline, or how steep the sloped surface is. The smaller the angle of incline, the longer the sloped surface and the easier the job will be.

An inclined plane with a greater angle of incline will have a steeper and shorter sloped surface. It will take more force to move an object along this plane.

If an inclined plane's horizontal distance is greater, the distance to the top will also be greater. This means that it will take longer to move the object to the top of the slope than if the inclined plane's horizontal distance was shorter.

NATURALLY OCCURRING INCLINED PLANES

◄---►

Inclined planes were found in nature before people started using them in their inventions. Streams and rivers travel downhill, along the sloped paths of hills and mountains.

Mountains have some of the largest naturally occurring inclined planes on Earth. Animals walk up and down winding paths on the hills and mountains where they live. It takes longer for them to climb up and down winding paths than if they were to climb straight up or down the side of a hill or mountain. However, it takes much less effort.

MECHANICAL MARVELS

Plant leaves can work like inclined planes, too. Leaves collect rainwater and direct it down to the plant's roots.

As water flows down a river or stream, the force of **friction** slows it down.

MAKING WORK EASIER

Simple machines are **designed** to make work easier for us. People have used inclined planes throughout history to make building temples, **aqueducts**, and roads easier.

MECHANICAL MARVELS

Some scientists and historians believe the ancient Egyptians used inclined planes to move the stone blocks needed to build the pyramids.

The ancient Romans built aqueducts this way. They also used inclined planes in war. In AD 72, the Romans attacked the Jewish fort at Masada, in present-day Israel, for many months. The fort sat at the top of a high cliff. In order to reach the top of the cliff and enter the fort without having to climb straight up the cliff, the Romans built a long ramp along the cliff's face.

Aqueducts are man-made **structures** for carrying water. These structures made it possible for farmlands and cities to get water from far away.

WORKING WITH INCLINED PLANES

←- →

We can find inclined planes in many workplaces. Trucks carrying food, furniture, and many other things in large amounts often have their own temporary ramps. These ramps can be set against the back of the truck to wheel or slide goods in and out.

Wood screws are actually tiny inclined planes. The thread of metal that circles the screw is like a winding road that climbs around a mountain. Wood screws cut through wood and allow us to fasten things together with less effort.

The rear bed of a dump truck also works like an inclined plane when the driver lifts one end of the bed and dumps out whatever is inside.

WOOD SCREW

INCLINED PLANES AROUND US

Inclined planes are all around us. You probably see and use inclined planes every day without even realizing it. At most street corners, there's a short incline at the curb. In front of your school or a post office or store, there may be a ramp to the door. These inclined planes make it easier for people with carts, strollers, wheelchairs, or walkers to enter and leave buildings.

Gutters and downspouts on houses and buildings are inclined planes that direct the flow of water to the ground.

Inclined planes also help people and objects travel down. The sides and bottom of a bathtub are inclined planes that direct the flow of water toward the drain.

THE HISTORY OF INCLINED PLANES

Even though humans have been using inclined planes to make work easier for thousands of years, these planes haven't always been considered simple machines. Many ancient mathematicians didn't group inclined planes with the other five simple machines because they don't move.

In 1586, Simon Stevin, a Flemish **engineer**, came up with a way to find an inclined plane's mechanical advantage using a string of beads. Around 1600, Italian scientist Galileo Galilei included the inclined plane in his writings about simple machines in *Le Meccaniche* (*On Mechanics*) because of how similar it is to other simple machines.

From 1874 to 1943, a **funicular** called the Incline carried people and goods to the top of Price Hill in Cincinnati, Ohio.

MODERN-DAY FUNICULAR IN SWITZERLAND

21

ENJOY THE RIDE

Inclined planes aren't only used to make work easier. They can also be used in sports. Ski hills, ski jumps, and skateboard ramps are all examples of inclined planes.

In winter sports such as luge, skeleton, and bobsled, people race very fast on sleds down icy inclined planes. People taking part in water sports such as wakeboarding and water skiing often use ramps for jumps to do tricks.

Slides at the playground and waterpark may be inclined planes that are just for fun. These don't make work easier, but they do make it possible to move down a slope quickly with the help of **gravity**!

GLOSSARY

aqueduct: A man-made channel constructed to move water from one place to another.

design: To create the plan for something.

determine: To officially decide something.

device: A tool used for a certain purpose.

engineer: Someone who plans and builds machines.

friction: The force that causes one moving object to slow down when it's touching another.

funicular: A railway going up and down a mountain that carries people in cars pulled by a moving cable.

gravity: The force that causes things to be drawn toward the center of Earth.

horizontal: Positioned from side to side rather than up and down.

permanent: Lasting for a very long time or forever.

structure: A building or other object that is constructed.

temporary: Lasting for a short amount of time.

vertical: Positioned up and down rather than from side to side.

INDEX

WEBSITES

Due to the changing nature of Internet links, PowerKids Press has developed an online list of websites related to the subject of this book. This site is updated regularly. Please use this link to access the list: www.powerkidslinks.com/dwsm/planes

24